Yacine BENALI

Polymer concretes and mortars

Yacine BENALI

Polymer concretes and mortars

Manufacturing and installation

ScienciaScripts

Imprint

Cover image: www.ingimage.com

This book is a translation from the original published under ISBN 978-620-6-71956-4.

Publisher:
Sciencia Scripts
is a trademark of
Dodo Books Indian Ocean Ltd. and OmniScriptum S.R.L publishing group

120 High Road, East Finchley, London, N2 9ED, United Kingdom
Str. Armeneasca 28/1, office 1, Chisinau MD-2012, Republic of Moldova, Europe
Printed at: see last page
ISBN: 978-620-7-99083-2

TABLE OF CONTENTS

INTRODUCTION 2

A HISTORY OF POLYMERS IN CONCRETE 3

CLASSES OF POLYMER CONCRETE AND MORTAR USED IN CONSTRUCTION 6

CONCLUSIONS 36

REFERENCE 37

INTRODUCTION

Portland cement-based concretes and mortars are among the most widely used construction materials in the world (1 m^3 per year per inhabitant) (Elalaoui, 2012), especially when combined with steel.

The mechanical qualities, relatively low cost, availability and ease of manufacture of Portland cement concrete and mortar are the most obvious advantages of these hydraulic composites, and have led to the hegemonic use of these materials in construction. However, low flexural strength, poor ductility, sensitivity to freeze-thaw effects and poor chemical resistance, and to some extent long-term durability, represent their most serious limitations (Afridi, 1995; Bahranifard, Vosoughi and Shariati, 2022). To overcome some of these shortcomings and to meet new needs for multifunctional materials with high performance, researchers have come up with composite materials, incorporating polymers instead of or in combination with Portland cement (polymer concretes or mortars).

The main objective of this document is to indicate and clarify current good practice recommended for the use of polymers in concrete and mortar. The document covers the different classes of these materials, their components and their potential applications in civil engineering. In addition, a bibliographical summary will be devoted to the main constituents, mixing, its mechanism, the formation of the co-matrix and the various interactions that govern the evolution of latex-modified mortars over time.

A HISTORY OF POLYMERS IN CONCRETE

The idea of using an organic polymer as a building material dates back thousands of years. Much evidence has shown that people in the early history of civilisation combined natural polymers with inorganic aggregates to produce durable, high-strength mixtures (Benali, 2018).

In the fourth millennium BC, the clay brick walls of Babylon were built using asphalt mortars which is considered to be a natural polymer (Elalaoui, 2012; Bothra and Ghugal, 2015). In addition, the temple foundations of Ur- Nina (king of Lagash), in the city of Kish in Iraq, are built from mortars that consist of 25% to 35% bitumen another natural polymer (Bothra and Ghugal, 2015).

Also, the walls of Jericho on the Jordan were also built using a bituminous polymer in around 2500-2100 BC, also bituminous mortars have been identified in the construction of the Indus Valley cities of Mohenje-daro and Harappa in Pakistan around 3000 BC and near the Tigris in Argentina in 1300 BC (Ribeiro, 2006; Bothra and Ghugal, 2015; Benali, 2018).

Many natural polymers, including albumin, blood, rice paste, and others have been used in ancient mortars (Bothra and Ghugal, 2015).

It is also believed that, as early as the 2nd century BC, glutinous rice paste with lime mortar was used to build China's Great Wall (Ribeiro, 2006; Benali, 2018). One of the seven wonders of the world.

The first synthetic polymers were developed in the early 20th century (Ribeiro, 2006; Bothra and Ghugal, 2015), but industrial production of these new materials did not see significant expression until the 1940s (Bothra and Ghugal, 2015). The first syntheses aimed to produce substitutes for two natural polymers: silk and rubber (natural rubber), whose sources of supply in Malaysia were cut off by the Japanese during the Second World War (Ribeiro, 2006).

However, since then, with increasing advances in polymer technology, hundreds of synthetic polymers have been produced with no equivalent in nature. Shortly after the end of the Second World War, the practice of incorporating synthetic polymers into concrete and mortar began (Ribeiro, 2006; Bothra and Ghugal, 2015). Apparently, the first indication of the use of natural polymers in cement concrete was in 1909, in the United States, when a patent was granted to L.H.Backland (Ribeiro, 2006; Ohama, 2011; Benali, 2018).

Thirteen years later, in France, another patent on cement mortars modified with natural polymers was granted to M.E.Varegyas (Ohama, 1995; Ribeiro, 2006).

In Britain, polymers in concrete known, in 1923 with the patent of L. Cresson with a rubber modify the paving materials of the road. The patent relates to paving materials with natural rubber latex where cement was used as filler (Bothra and Ghugal, 2015) .

In 1924, with the development of L. Lefebure the first patent with the current concept of polymer modification (rubber latex) was published, and later, in 1925, another patent concerning innovation on this product was issued to Kirkpatrik (Ribeiro, 2006; Bothra and Ghugal, 2015).

The use of polymers has been seen as a sign of progress and a modern attitude in construction and the polymer-concrete industry is proving very interesting, especially as these "new materials" are attracting renewed interest thanks to their remarkable qualities compared with conventional construction materials in the field of civil engineering (Benali, 2018).

So what are polymers and why are they used in concrete and mortar? The word polymer is a Greek word composed of "pollus" and "mero", meaning "many" and "part" respectively. The general concept of polymer composites from concrete, from a technical point of view, involves a process by which chemical products: monomers, oligomers, prepolymers, polymers are introduced into a concrete or mortar mixture, then they are subjected to a chemical reaction -

polymerisation or polycondensation (covalent linking of several chemical units or unitary motifs) by thermo-catalytic or other systems.Incorporating polymers into concretes and mortars obviously produces materials similar to ordinary cementitious concretes, but with superior characteristics (Bothra and Ghugal, 2015). The cementitious matrix can be replaced in part or in whole, by an organic matrix "a resin". These are generally classified into three classes.

CLASSES OF POLYMER CONCRETE AND MORTAR USED IN CONSTRUCTION

The search for methods and means of increasing the density, resistance to chemical attack and durability of reinforced or unreinforced concrete has led to the formulation of different groups of concrete with additives or based on polymers, known as polymer concretes.

The best-known polymer concrete systems are generally classified according to process technology. Figure 1 summarises the concrete systems used in construction.

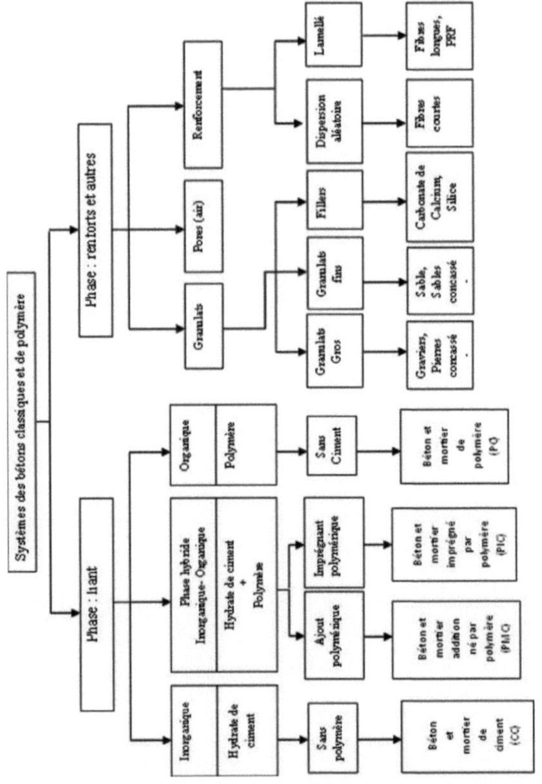

Figure 1: Classification of concrete systems used in construction

The figure shows that the use of polymers in concrete results in :

- The chemical nature of these basic components.
- How it is made (with water - without water - with cement or without water and cement)
- The percentage of polymer introduced into the mixture ratio (W/C).
- The nature of the polymer "form" (monomers - liquid resin - redispersible powder - solution).

Polymer-impregnated concrete and mortar (BIP/MIP)

BIPs and MIPs were the first composites to be developed. They have been widely known since the late 1960s to 1970s (Fowler, 1999; Ohama, 2011). These systems are composites formed by injection of a low viscosity monomer (Fowler, 1999; Knapen, 2007; Elalaoui, 2012), typically methyl methacrylate MMA (ACI_Committee_548, 1997; Fowler, 1999) in liquid or gaseous form into the pores of Portland cement concrete (or mortar) after curing (Fowler, 1999; Elalaoui, 2012); although liquid type monomers are more easily adapted to the impregnation of cured concrete (or mortar) (ACI_Committee_548, 1997). In general, almost any shape, size, configuration, quality of Portland cement concrete (or mortar) can be impregnated, provided that the monomer has access to the void space within the concrete (or mortar). A significant part of this space is generally obtained by the elimination of free water from the pores by simple drying (ACI_Committee_548, 1997). It is clear that the filling of the available space in the pore network of the concrete (or mortar) with monomers determines whether this composite is partially or totally impregnated. According to the guide (ACI_Committee_548, 1997), total impregnation implies that approximately 85% of the empty space available after drying is filled. After impregnation, the concrete or mortar containing the desired amount of monomer is subjected to a treatment to convert the monomer into polymer (polymerisation).

The two most commonly used methods for polymerisation in impregnated systems are called thermocatalytic and promoted-catalytic. A third method, involving ionising radiation, is used less frequently (ACI_Committee_548, 1997; Fowler, 1999).

A description of the process for the BIPS, using the thermocatalytic polymerisation is shown in Figure 2.

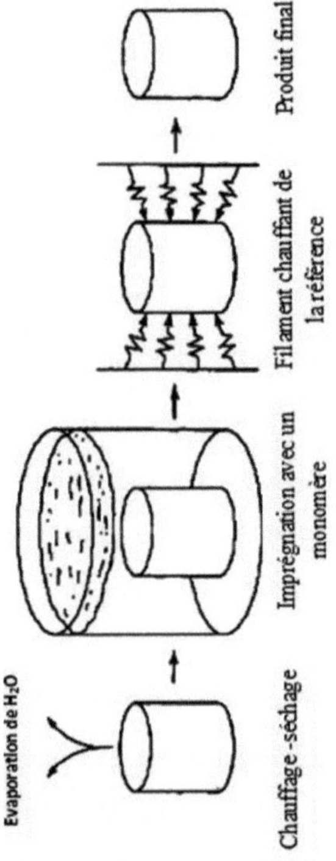

Figure 2. Diagram of the BIP production method (ACI_Committee_548, 1997)

PROPERTIES OF IMPREGNATED SYSTEMS

Generally, impregnation of concrete (or mortar) positively affects several important properties, including tensile, flexural and compressive strengths (Fowler, 1999; Elalaoui, 2012). In most cases, these new polymer-impregnated materials also have a compressive strength equal to four or five times that of conventional concrete (ACI_Committee_548, 2009). Also an improvement can be reported on Young's modulus, abrasion resistance, resistance to penetration and damage by water, acids, salts, and freeze/thaw cycle resistance (Fowler, 1999; Elalaoui, 2012; Czarnecki, Őzkul and Wang, 2013), see Table 1.

Table 1. Durability of polymer-impregnated concrete (Elalaoui, 2012).

		Control concrete	Polymer-impregnated concrete (MMA)	Polymer-impregnated concrete (Styrene)
Freeze-thaw	Number of cycles	740	3650	5440
	Mass loss (%)	25	2	2
Sulphate attack	Expansion (%)	0.466	0.006	0.03
	Number of days	480	No	690
Acid resistance (15% HCl)	Mass loss (%)	27	9	12
	Number of days	105	805	805
Wear resistance	Depth depth (mm)	1.25	0.38	0.93
	Mass loss (g)	14	4	6

APPLICATIONS OF IMPREGNATED SYSTEMS

Thanks to its exceptional properties, it can be used in a wide range of applications, including :

• Preserving monuments.

• Rehabilitation of old and run-down buildings.

• The manufacture of bridge decks, pipes and conduits for aggressive fluids.

• They have also been used to make tiles, beams, slabs, pavements, kerbs and tunnel linings (Fowler, 1999; ACI_Committee_548, 2009).

• What's more, its systems are widely used to make ponds and water covers.

• They are also used to make thin prefabricated panels and to repair damaged road surfaces (Kardon, 1997).

Limits of impregnated systems

The main disadvantage of impregnated systems (BIP and MIP) is their cost, which is not economically viable. Currently, the monomers used for impregnation are expensive and the manufacturing process is more complicated than for conventional products (Knapen, 2007; Ohama, 2011; Elalaoui, 2012). Also, partial impregnation does not make the concrete completely waterproof, and exposure to very aggressive agents, such as sulphuric acid, can be detrimental (the concrete or mortar can be slowly attacked).In addition, the impregnation technique requires the application of heat to dry the surface of the concrete, which can cause cracking and lead to a reduction in strength. However, the impregnation process takes a long time (around two days for partial impregnation) (ACI_Committee_548, 2009).

Concrete or polymer mortar (BP/MP)

They belong to the category of organic matrix materials, hence its designation of organic matrix concrete (or mortar) (Haidar, 2011; Elalaoui, 2012). It is also known as synthetic resin concrete (or mortar) or plastic resin concrete (or mortar) or simply as resin or polymer concrete (or mortar) (Ribeiro, 2006; Haidar, 2011; Benali, 2018). This system came into use in the late 1950s and early 1960s (Ohama, 2011), although demand was very limited. It was first used in the construction industry in the United States, then in 1958 for the innovative manufacture of prefabricated wall panels and artificial marble (Fowler, 1999). In Europe, it was used as a repair material as early as 1961 (Ribeiro, 2006). However, it was not until the late 60s and early 70s, after the development of BIPs, that BP became better known. These composites are generally formed by a granular skeleton and a polymer binder (Haidar, 2011). They do not contain a hydrated cement phase (Salbin, 1996; Kardon, 1997; Czarnecki, 2007; Zabihi and Ozkul, 2015), although Portland cement can be used as a filler "without the water" (Ribeiro, 2006). When the reinforcements are formed by sand or smaller inclusions, the composite is said to be a resin or polymer mortar (Haidar, 2011; Elalaoui, 2012).

The polymers used for this system are sometimes thermoplastics, but in most cases they are thermosets, see Figure 3.

a) MMA - methyl methacrylate;

b) TMPTMA - trimethylolpropane trimethacrylate;

c) HMWM -High Molecular Weight Methacrylate; d) UMA -Urethane Methacrylate.

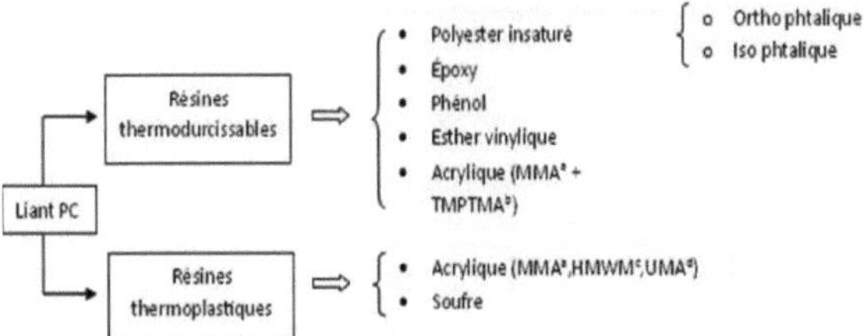

Figure 3. Commercially available polymers used as binders in BP/MP systems (Ribeiro, 2006)

Properties of concrete or polymer mortar composites

The total replacement of the cementitious matrix in the case of ordinary concretes or mortars by a polymeric binder is justified by the improvement of its behaviour against various types of mechanical, chemical or other aggression(Elalaoui, 2012; Czarnecki, Őzkul and Wang, 2013).

This type of material offers a number of advantages, including :

- Good resistance to chemical and corrosive agents;
- Low water permeability and good resistance to freeze-thaw cycles;
- Low coefficient of thermal expansion ;
- Rapid curing at ambient temperatures (-18 to 40°C) ;
- Good adhesion to aggregates and old concrete;
- Better mechanical resistance than hydraulic concrete, especially in tension, see table 2;
- Good abrasion resistance;
- Excellent durability at a reasonable cost and low weight;
- As a result, they have a reduced modulus of elasticity and can be used with thin layers for decorative and architectural finishes;
- What's more, they have good creep resistance and are highly resistant to UV

rays. due to the very low polymer content and inert fillers.

Table 2. Typical properties of common polymer concretes and Portland cement concrete (Elalaoui, 2012).

Material	Density (kg/l)	Absorption water content (%)	Rc (MPa)	Rt (MPa)	Rf (MPa)	E (GPa)
Polymethyl methacrylate	2-2.4	0.05-0.6	70-210	9-11	30-35	35-40
Polyester	2-2.4	0.3-1	50-150	8-25	15-45	20-40
Epoxy	2-2.4	0.02-1	50-150	8-25	15-50	20-40
Furane	1.6 - 1.7	0.2	48-64	7-8	-	-
Portland cement concrete	1.9-2.5	5-8	13-35	1.5-3.5	2-8	20-30

A pplications of concrete or polymer mortar composites Because of its many advantages, polymer concrete or mortar became a dominant material in the construction industry in Japan and Europe during the 1970s and in the 1980s in the United States (Ohama, 2011; Elalaoui, 2012).

They have been used in applications where there is a need for durability, such as in the manufacture of drainage and piping systems (manholes, acid and hazardous waste containment tanks, etc.).

They have also been used for surface repair coatings and prefabricated products. Table 3 summarises the major applications of these composites.

Table 3. Typical BP applications (Ribeiro, 2006).

Function/ Application	End uses
Floors and pavements	Floors in houses, sheds, schools, hospitals, offices, shops, toilets, corridors, sports halls, factories, industrial plants, staircases, garages, platforms, etc. railway lines, airport runways, etc.
Coatings waterproofing	Concrete roofs, water covers, pits and swimming pools, silos, etc.
Decorative coatings	Wall coverings, coating materials for textured building finishes, surface preparation equipment for coatings, etc.
Anti-corrosion coatings	Effluent from drains, grout for acid-resistant tiles, floors for chemical and pharmaceutical laboratories, hot baths for springs, coatings, etc. anti-corrosion for terraces with steel roofs, etc.
Bridge cladding	Decks, internal and external ship decks, the terraces of footbridges, the floors of trains and underground trains, etc.

Limits of concrete or polymer mortar composites

These systems have a number of limitations. The cost of the raw material (mainly the binder) is higher (up to 8 times) than that of hydraulic binders (Elalaoui, 2012). According to (Adams, Browne and French, 1975), the cost of epoxy concrete is 125 times that of ordinary reinforced concrete for a 6.5-fold increase in flexural strength.

This additional cost can be offset in some applications by the reduction in labour costs and the lower price of polymer-based products, which are only 10 to 25% more expensive than hydraulic binders (Elalaoui, 2012).

This saving is achievable thanks to a number of factors, such as material savings (up to 50%) by increasing product dimensions thanks to the high strength of resinous concrete, reduced transport costs thanks to smaller volumes, and a considerable reduction in capital expenditure during manufacture and during handling and storage operations as a result of This reduces the space required for production and storage of finished products. In addition, these concretes are characterised by a bad smell and toxicity from the binding part of the material (the resin and hardener) during mixing and placing. The behaviour of these composites at high temperatures and in the presence of fire also hampers their development, especially when used as a cladding material or for interior decoration. Since resins are organic substances that do not resist heat well, their prolonged exposure to high temperatures is not recommended, as it leads to their degradation, which eventually results in a loss of mechanical strength (Adams, Browne and French, 1975; Kardon, 1997; Elalaoui, 2012).

Polymer-modified concrete or mortar (BMP/MMP)

This material, which has a polymer content of more than (5%) compared with cement, came into use in the early 1950s and in the 1970s. It has become one of the dominant construction materials in advanced countries (Ohama, 2011). It is defined as a mixture of binder (Portland cement) and aggregate combined at the time of mixing with an organic polymer usually as an admixture (Kardon, 1997; ACI_Committee_548, 2003; Czarnecki, Őzkul and Wang, 2013). When cement hydrates and polymer coalescence occur, a co-matrix is driven through the concrete (or mortar). As a result, a hybrid matrix is obtained, formed by this co-matrix, which leads to an improvement in the final product. According to (Ohama, 1995), modified mortars are more widely used than concrete in terms of the balance between performance and cost. Several types of polymer are used to make polymer-modified concrete or mortar. The most commonly used are in the form of an aqueous dispersion known as a latex (Kardon, 1997; Haidar, 2011). There are also redispersible polymers, water-soluble polymers and liquid resins (Knapen, 2007; Zabihi and Ozkul, 2015), see Figure 4.

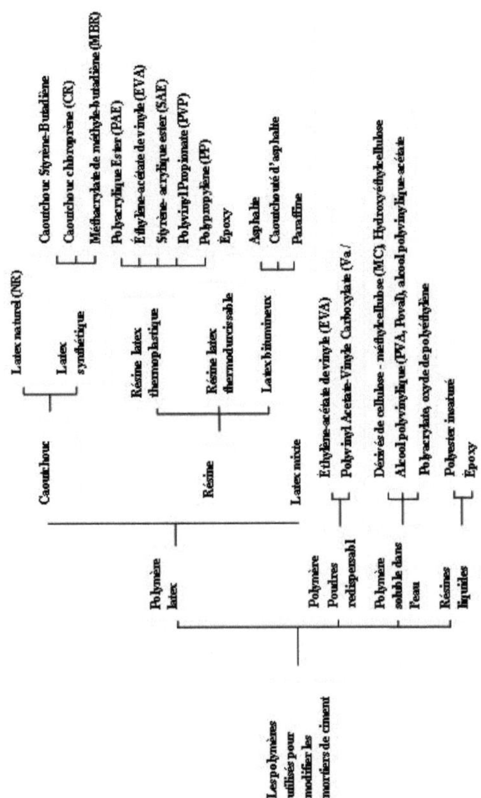

Figure 4. Polymers for modifying mortar and cement concrete (Ohama, 1998; Knapen, 2007).

REDISPENSABLE POWDER-MODIFIED MORTARS

The principle of powder modification is almost the same as that of latex modification, except that in place of the aqueous solution (the latex), there are redispensable polymer powders (Ohama, 1995). These powders are mainly used for the manufacture of mixes whose constituents are: cements and aggregates (generally < 6 mm in size), water and antifoam agents if the powder does not contain them (ACI_Committee_548, 2003). After the addition of water, redispersible powders are re-emulsified in the cement mixture, and behave in the same way as latexes (Knapen, 2007). The dosing of polymer-modified cement mixes in redispersible powder form is similar to that of other polymer-modified systems, with the exception that the water is not supplied by the polymer. The P/C ratio by mass varies from 5 to 20% (ACI_Committee_548, 2003). The hardening of modified systems is similar to that of latex-modified systems (Ohama, 1995).

Manufacture of redispensable powders

The first redispensable powders were based on polyvinyl acetate. They were developed by (Max Ivanovits), a research chemist at Wacker AG, in 1952 (Baueregger, 2014). These powders are obtained by a two-stage process (Ohama, 1998). Firstly, the polymer latexes that are the raw materials are produced by emulsion polymerisation. These are then dried by spraying to obtain what are known as redispensable powders (ACI_Committee_548, 2003; Knapen, 2007; Baueregger, 2014). Before spray-drying, latexes are further formulated with ingredients such as bactericides, spray-drying aids and anti-foaming agents. Anti-blockers such as clay carbonate, silica and calcium are added to the polymer powder during or after spray drying to prevent the powders from setting during storage. At present, the only polymers available in the form of reusable powders are PAE, SA, VAE, VA-VEOVA and PVA (Ohama, 1995; ACI_Committee_548, 2003). The properties of mortars modified with redispensable powders The advantage of using re-dispersible powders is that the mix dosage is better controlled, with the dry ingredients usually dosed in the factory rather than on site, as is the case with latexes.As a general rule, the same properties reported for mortars after latex modification are also obtained for mortars after modification with redispensable powders (Ohama, 1995); only these properties are slightly reduced compared to those of latex of similar composition (ACI_Committee_548, 2003). Indeed, the addition of latexes to mortars tends to form a polymer film more easily and uniformly than the addition of redispensable powders (Afridi et al., 2003).

Applications of redispensable powder-modified systems

Redispensable powders are always more expensive than their latex equivalents because they are generally obtained by a two-stage process as mentioned above. As a result, powder-modified systems are used where cost is not a major concern and convenience is more important, such as in applications that require the use of small quantities. According to the guide (ACI_Committee_548, 2003), mortars modified with redispensable powders are used very effectively in :

- Ceramic tile adhesives and grouts ;
- For the construction of industrial floor coverings ;
- For repairing concrete structures and mortars;
- In addition, these powders are used to a limited extent in exterior insulation finishing systems.

Mortars modified with water-soluble polymers

This type of modified mortar uses water-soluble polymers instead of latexes and powders with very low polymer ratios, varying between 0.5 and 4%, depending on the type of polymer used (Knapen, 2007). These polymers can be added as powders or as aqueous solutions to cement mortars during mixing (Ohama, 1995, 1998). When added in powder form, it is advisable to dry mix the polymers with the overall cement mixes and then mix them with water. The literature shows that this type of modification is applied less frequently (Knapen, 2007), because there are few polymers that are soluble in water. In addition, their low solubility causes difficulties when using them as mortar modifiers. On the other hand, these polymers have the advantage of the absence of surfactants, which are used to keep the polymers in solution. The absence of surfactants facilitates the formation of the polymer film, and the properties of the materials can be better controlled. Consequently, low polymer contents are required (Knapen, 2007).

Frequently used water-soluble polymers

There are non-ionic polymers with an oxygen or nitrogen atom in the polymer backbone. Examples include polyethylene oxide (PEO) and polyethyleneimine (PEI). There are also water-soluble non-ionic polymers containing an acrylic group, such as polyacrylic acid (PAA) and polyacrylamide (PAAM). In addition, soluble polymers such as polyvinyl alcohol (PVA) and polyvinyl alcohol acetate (PVAA), which are frequently used to modify cement mortars, belong to the class of water-soluble non-ionic polymers containing a vinyl group. Finally, there are cellulose ethers, such as methyl cellulose (MC) and hydroxyethyl cellulose (HEC), and polyelectrolytes, such as polystyrene sulphate (PSS) (Knapen, 2007).Another classification is given by (Knapen, 2007): natural, semi-synthetic and synthetic polymers. Natural polymers include starches, etc. Cellulose derivatives such as hydroxyethylcellulose and methylcellulose belong to the semi-synthetic groups. Finally, synthetic polymers include ethylene-based polymers such as polyethylene oxide, and vinyl-based polymers such as polyvinyl alcohol and polyvinyl alcohol acetate. Acrylates such as calcium acrylate and polyvinyl acrylate are also used. magnesium, which are added as monomers during mixing, are also water-soluble polymers (Ohama, 1998). No special cure is required for systems modified with soluble polymers (Ohama, 1995). Properties of mortars modified with water-soluble polymers According to Ohama (Ohama, 1995), systems modified with soluble polymers exhibit remarkable water retention, air entrainment and plasticising effects. This can make a major contribution to improving workability and preventing the phenomenon of "The water retention capacity of these polymers also contributes to superior adhesion to porous substrates such as ceramic tiles and cementitious materials (Jenni et al., 2005; Knapen, 2007). In addition, the water retention capacity of these polymers can also contribute to superior adhesion to various porous substrates such as ceramic tiles and cementitious materials (Jenni et al., 2005; Knapen, 2007). In addition, this water retention capacity often results in increased viscosity, which tends to reduce free water and the tendency for segregation in the cement paste (Knapen, 2007). In addition, homogeneity and better dispersion of the carbon and steel fibres are obtained in the modified cement paste (Knapen, 2007). On the other hand, water-soluble polymers tend to decrease compressive strength due to the increase in entrained air (Ohama, 1995; Fu and Chung, 1996a); while an increase in tensile strength is found for some applications (Fu and Chung, 1996a). On the other hand, the bond strength between cement paste and aggregates (Knapen and Gemert, 2015), between cement paste and steel fibres (Knapen and Gemert, 2015), and between cement paste and carbon fibres (Fu, Lu and Chung, 1996), is improved. In their work (Fu and Chung, 1996a) investigated the effect of incorporating methyl cellulose into mortars. They found that tensile strength was increased by up to 72%, fracture ductility by up to 620%, while compressive strength was decreased by up to 30% and compressive ductility by up to 34%. On the other hand, the same authors in another study (Fu and Chung, 1996b) found that the addition of 0.4% methylcellulose and 20% latex to cement paste gave a similar and significant increase in adhesion between stainless steel fibres and cement paste.

In addition, improving the interface between aggregates and cement paste can also lead to a decrease in both permeability and crack formation (Knapen and Gemert, 2015). These polymers, even if added in small quantities, have the ability to form a polymer film in the cementitious mix, as previously mentioned. This trend is cited by a few authors in the literature (Knapen and Gemert, 2015). The latter studied the presence of the polymer film in the hardened structure of cement mortars modified with polyvinyl alcohol acetate, methyl cellulose and hydroxyethyl cellulose. The results found proved the presence of polymer films in mortars modified with 1% polyvinyl alcohol acetate and methyl cellulose. The same trends were reported by (Jenni et al., 2005). Generally speaking, the aim of modifying water-soluble polymers is to improve the fresh properties of the mortar, while the impact on mechanical properties is considered to be a secondary effect (Knapen, 2007).

Applications of mortars modified with water-soluble polymers

These polymers are often used as additives in the manufacture of adhesive mortars for ceramic tiles and self-levelling screeds (Knapen, 2007). They are also frequently used as water retention agents, so-called rheology modifiers or viscosity modifiers (Knapen, 2007). They are also used to repair damaged structures. These systems also have properties that make them suitable for underwater applications. Soluble polymers are also used to manufacture cements (MDF), which have very high tensile strength. These systems are generally made at very low water-cement ratios ranging from 0.08 to 0.2 (Donatello, Tyrer and Cheeseman, 2009), at high pressures and temperatures. The most common water-soluble polymers used to make these cements are polyvinyl alcohol or polyvinyl alcohol-ethyl acetate (Chetrashekhar, Cooper and Shafer, 1989; Zhang, 2014), polyacrylamide and hydroxypropylmethylcellulose (Knapen, 2007).

Mortars modified with liquid resins

These are mortars modified with viscous polymers such as epoxy resin, unsaturated polyester resin or polyurethane resin (Ohama, 1998). The mixing of this type of composite is similar to that of latex-modified systems (Ohama, 1995, 1998; Li et al., 2022). or curing agents to form a network structure (two-component polymers). Mortars modified with liquid resins are usually obtained by first mixing The mixture consists of cement, aggregates and half the mixing water, followed by the hardener pre-mixed with the resin and the rest of the mixing water. The resin/hardener content is practically the same as or higher than that of modified latex systems; the most common vary between 15 and 20% (Ohama, 1995, 1998). However, according to Ohama (1995), epoxy resin hardening can occur in the alkaline environment of cement mortars without hardener. These authors carried out a study on mortars modified with epoxy resins without hardener. They observed that the mortars were made successfully, and had superior properties compared with mortars modified with conventional epoxy resins (with hardener). In addition, they concluded that a polymer-cement ratio ranging from 5 to 10% is considered optimal for the preparation of these new mortars. When it comes to curing, these systems have the advantage of curing in damp or wet conditions.

The properties of mortars modified with liquid resins

Unlike soluble polymers, few studies are devoted to the effect of liquid resins on the fresh properties of mortars. The majority of studies that exist in the literature mention the hardened properties of mortars modified by liquid resins.

In this type of mortar, polymerisation is initiated in the presence of water to form a polymer phase, while cement hydration occurs at the same time. The degree of hydration probably decreases as the cement particles are covered by the resin particles, reducing the contact between the cement and the water.

Consequently, the addition of epoxy resin delays the hardening of pastes and mortars. After curing, a co-matrix phase is formed with a combined polymer network structure and a phase of cement hydrates and aggregates that interpenetrate, binding them strongly. As a result, the strength and other properties of the mortar are improved in much the same way as those of latex-modified systems (Ohama, 1998). This means high strength and adhesion, low permeability, and improved chemical resistance just like latex-modified systems (Ohama, 1995). In their studies (Aggarwal, Thapliyal and Karade, 2007) aimed to compare the properties of cement mortars modified with epoxy emulsions and those modified with acrylic latexes. They found that for a similar (W/C) ratio, epoxy-modified mortars showed relatively better mechanical properties and durability than acrylic emulsions.

Applications of mortars modified with liquid resins

According to (Ohama, 1995, 1998), liquid polymers are less widely used as additives in Portland cement mortars than other additives such as latexes, redispersible polymer powders and water-soluble polymers. These resins are produced to make polymer concretes and mortars without hydrated cement.

Latex-modified mortars

These mortars contain a latex-type polymer plus the same ingredients as conventional mortars (cement, aggregates, water and other additives if used). This type of mortar will be discussed in detail in this book. As previously mentioned, these mortars are prepared by adding to the conventional mortar components (cement, sand and water), a polymer in the form of an aqueous dispersion (latex) (Salbin, 1996; Ngassam, 2013), often used as a water-reducing plasticising admixture (Dhaini, 2014). The latex polymer/cement ratio varies from 5 to 20% polymer solids by mass of cement in the mix (Ohama, 1995; ACI_Committee_548, 2003; Zohhadi, 2014).The polymer latexes are first mixed with the mixing water, then directly added to the cement and then mixed with the aggregates; this is the classic method recommended by Ohama (Ohama, 1995) and the ASTM C 1439 standard (ASTM_C1439, 1999), see table 4. Mixing speed and time must be carefully chosen to avoid trapping unnecessary air. The air bubbles that form when cement and aggregates are mixed with an aqueous polymer solution are not easy to remove, as they tend to be stabilised by the polymers, resulting in reduced strength (Ngassam, 2013; Ukrainczyk and Rogina, 2013), even though this phenomenon can be beneficial in improving the workability and durability of mortars.

Table 4. Mixing according to ASTM C 1439 (1re Method).

Operations	Duration of operations (s)	Mixer condition
Introduction of water, the latex and anti-foaming agent.		Stop
Introduction of cement		
	30	Slow speed
Introduction of sand	30	
	30	Average speed
Scraping the tank	15	Stop
	75	
	60	Average speed

Some researchers (Salbin, 1996; Ukrainczyk and Rogina, 2013) suggest using antifoam agents in the mixture, in cases where these agents are not contained in the latex. Others (Kim and Robertson, 1997; Barluenga and Herna'ndez-Olivares, 2004; Li, Zhong and Zhang, 2012), propose using an alternative approach to reduce the formation of air bubbles in modified mortars, which is to pre-wet the cement and sand with water, then add the latex (see Table 5).

Table 5. Mixing using pre-wetting (2nd method).

Operations	Duration of operations	Mixer condition
Introduction of water		Stop
Introduction of cement		
	30s	Slow speed
Introduction of sand	30s	
	30s	Average speed
Scraping the tank	15s	Stop
	75s	
	60s	Average speed
	120s	Vibration
Introduction of latex	180s	Low + speed vibration

Description of the main constituents of latex-modified mortars

• **Cement :**

The most commonly used are portland cements (Dhaini, 2014), which are mineral powders that harden when in contact with water (Goto, 2006; Ngassam, 2013).

a) Composition and manufacturing process of Portland cement :

It is obtained by grinding clinker with approximately (3 to 5%) gypsum (calcium sulphate), which regulates its setting (Belkhodja, 2013; Dhaini, 2014). Clinker is produced after high-temperature firing (1450°C) of a mixture of calcium carbonate ~ 80%, and 20% clay composed of: Silica ~ 15% alumina ~ 3%, Ferric oxide ~ 3% (Belkhodja, 2013; Baueregger, 2014).

b) Hydration of cement :

The cement phases hydrate, at different rates, following an exothermic process to form a cohesive material (Dhaini, 2014).

➢ Hydration of silicates :

Portland cements react with water to form hydrated calcium silicate (C-S-H) and calcium hydroxide (CH), known as portlandite (Ngassam, 2013). The reactions taking place are as follows (1, 2):

$2\ C3S + 8\ H \rightarrow C3S2H5 + 3\ CH$ (1)

$2\ C2S + 6\ H \rightarrow C3S2H5 + CH$ (2)

Hydration of C3S alite is rapid (Ngassam, 2013). It begins directly after the first few hours of mixing (Ngassam, 2013). It is therefore this phase that is responsible for the setting of the cement paste, cohesion, and the evolution of the cement's mechanical properties at a young age (Ngassam, 2013; Dhaini, 2014). On the other hand, the hydration of C2S belite is much slower (Ngassam, 2013). Their contribution to the cement's mechanical strength becomes significant after one week (Ngassam, 2013).

➢ Hydration of aluminates :

C3As are highly reactive in the absence of sulphates (Ngassam, 2013). They cause the cement paste to stiffen (flash setting) and prevent it from setting (Dhaini, 2014). This is why gypsum (CSH2) is used in cement manufacture (Ngassam, 2013). In the presence of water, aluminates (C3A and C4AF) react with gypsum to form hydrated calcium trisulfoaluminate, referred to as ettringite (C6AS3H31) (Ngassam, 2013; Dhaini, 2014). The reaction taking place is as follows (3).

$C3A + 3\ CSH2 + 25\ H \rightarrow C6AS3H31$ (3)

The remaining C3A aluminates react with some of the ettringite to give calcium monosulphoalumininate (4) and with water to give hydrated aluminates (5).

$C3A + 14\ H + C6AS3H31 \rightarrow 3C4AS3H15$ (4)

$2\ C3A + 21\ H \rightarrow C4AH13 + C2AH8$ (5)

Cement also contains small quantities (less than 1%) of alkalis in the form of oxides which dissolve completely on contact with water, releasing hydroxyl ions (6, 7).

$Na_2O + H_2O \rightarrow 2\ Na^+ + 2\ OH^-$ (6)

$K_2O + H_2O \rightarrow 2\ K^+ + 2\ OH^-$(7)

An equilibrium is also established in the medium between the portlandite and the calcium ions (8): $Ca(OH)_2 \leftrightarrow Ca^{2+} + 2\ OH^-$ (8)

c) Sustainable development and environmental impact of portland cements :

Global warming linked to emissions of greenhouse gases such as carbon dioxide (CO_2) has been observed since the middle of the 20th century. Controlling these emissions has therefore become a major global challenge. This problem involves all sectors of activity, particularly those relating to energy management, natural resources, raw materials and transport (Bur, 2012).

In the field of civil engineering, taking sustainable development requirements into account means reducing the environmental impact of structures throughout their life cycle, while maintaining their quality of use (functionality and performance). Analysis of the life cycle of cement structures shows that, of the various constituents used, cement and particularly cement production is the main source of environmental impact (Bur, 2012). Not only does it consume limestone, clay, marl and fuel, but it is also responsible for the majority of greenhouse gas emissions. Clinker production requires a large quantity of energy (of the order of 4 GJ/t) (Bur, 2012). It is also one of the biggest emitters of carbon dioxide, emitting 5 to 10% of global CO_2 emissions into the atmosphere (Deventer, Provis and Duxson, 2012; Atsonios et al., 2015). As a general rule, one (01) kilogram of clinker generates around 0.9 kg of CO_2 (Hasanbeigi, Price and Lin, 2012; Brien and Mahboub, 2013).The difference between this and other industries is that fuel combustion is not the only dominant factor in CO_2 emissions (Hasanbeigi, Price and Lin, 2012). More than half of the CO_2 emissions produced during the cement manufacturing process come from the calcination of raw materials, while the remainder comes from the combustion of fuels to provide a temperature of 1450°C, at which clinkerisation takes place (Hasanbeigi, Price and Lin, 2012; Hot, 2013). In addition, quarries, transport and production sites are all potential sources of noise, vibration, dust and various polluting gases such as nitrogen oxides (NOx) or sulphur oxides (SOx), which can have an impact on public health. Reducing CO2 emissions in cement production now involves reducing the proportion linked to the decarbonation of raw materials, and the most effective way at present seems to be to reduce the quantity of cement by using substitute materials (Bur, 2012). In particular, the use of industrial by-products added directly to clinker or in high filler as substitute materials in cement mixes. The most commonly used products are fly ash, silica fume and blast furnace slag (Bur, 2012).

- **Latex :**

Latexes are hydrophobic spherical polymer particles suspended in an aqueous phase (Goto, 2006; Dhaini, 2014). They are also defined as colloidal suspensions of polymers stabilised in water (Belkhodja, 2013; Ngassam, 2013). Latexes first appeared in the 1930s (Dhaini, 2014). The average particle size varies from 50 to 5000 nm (ACI_Committee_548, 2003; Sivakumar, 2011).

a) The synthetic latex manufacturing process :

Most latexes are manufactured using a process known as emulsion polymerisation (Thickett and Gilbert, 2007; Kapil Soni and Joshi, 2014). This is a highly successful process from both a technical and environmental point of view, thanks to the use of water as a solvent and also the negligible quantity of volatile organic compounds (VOCs) released during their preparation and application (Amaral, 2004; Nguyen, 2008; Erdmenger et al., 2010; Pragliola, Vita and Longo, 2013). These are petrochemical monomers manufactured from crude oil (Elalaoui, 2012). However, they can also be produced from coal, natural gas, wood or vegetable substances (Elalaoui, 2012). After the petroleum distillation phase, naphthas are broken down into light molecules known as "monomers" by steam cracking at a temperature of 850°C and in the presence of steam. These monomers are first dispersed in the form of droplets in an aqueous phase (Benyahia, 2009; Youssef, 2012). To these droplets is added a water-soluble or water-soluble initiator (ACI_Committee_548, 2003; Benyahia, 2009; Youssef, 2012) and a surfactant, often of the non-ionic (fresh) type for use with Portland cements (Salbin, 1996; ACI_Committee_548, 2003). The role of the surfactant is to stabilise the dispersion of the monomer droplets and polymer particles (Benyahia, 2009), since without stabilisation, the particles tend to aggregate by flocculation, under the effect of Brownian motion (Ngassam, 2013). Sometimes pH buffers are added, as well as antifoam agents and chain transfer agents to control the molar masses, which are often excessive (Salbin, 1996; ACI_Committee_548, 2003).

b) The main types of latex used to modify mortars:

A wide variety of latex types have been studied for use in modified mortars (Ohama, 1995; Salbin, 1996; Nicot, 2008; Ngassam, 2013); but the main types in use today are: Acrylic polymers and copolymers (PAE), styrene-acrylic copolymers (SA), styrene-butadiene copolymers (SBR), vinyl acetate copolymers (VAC), and vinyl acetate homopolymers (PVA). The main vinyl acetate copolymers are ethylene (VAE) and versatic acid vinyl ester (VA/VEOVA), although acrylic acetate copolymers are also used to some extent.

Recently, Styrene Butadiene Rubber (SBR) and Acrylic Polymers (SA and PAE) are the main types in use (Benali and Ghomari, 2017). These latexes account for 37% (SBR) and 30% (SA and PAE) respectively of the total water-based synthetic latexes manufactured industrially (Benali, 2018). They have been used for a long time to modify mortars and appeared to be very good (Benali and Ghomari, 2017). In what follows, we first present the process for forming the polymer film and the co-matrix, and then briefly describe the influence of latexes on the mechanical properties and durability of cement matrices.

c) Filmification of latex :

For applications such as coatings or mortars, latex polymers undergo a drying step (Baueregger, 2014). Due to the loss of water through evaporation and then hydration of the cement, a homogeneous polymer film can form above a threshold temperature called the minimum film forming temperature (TMFF) (Kardon, 1997; Goto, 2006; Ngassam, 2013). Below this temperature, the polymer cannot form this continuous film (Kardon, 1997; Nicot, 2008; Belkhodja, 2013).Another temperature used to characterise latexes close to the TMFF temperature is the glass transition temperature (Tg) (Nicot, 2008). This is the temperature at which the chains become mobile (Goto, 2006). It corresponds to the temperature at which the polymer's behaviour changes from the glassy, hard and relatively brittle state to a softer "rubber" state (Kardon, 1997; Goto, 2006). The hardness of the modifying polymer is also linked to this temperature Tv. In general, the higher the value of Tv, the harder the polymer and the higher the resulting compressive strength. On the other hand, when Tv is low, the resulting system has low permeability (ACI_Committee_548, 2003). As for this film, recent research work (Baueregger, 2014), describes the mechanism of its formation (see figure 5). Initially, the loss of water from the latex (stage 1) increases the concentration of polymer in the mixture, subsequently generating the formation of a stack of polymer in contact (stage 2). This gradually deforms (stage 3), which occurs at temperatures above the minimum film-forming temperature (MFT) so that the inter-particle spaces disappear as a hexagonal structure is formed (Baueregger, 2014). Coalescence of the latex particles then occurs, when the ambient temperature exceeds the glass transition temperature of the polymer (stage 4) (Baueregger, 2014). Here, the interfaces between the latex particles break down and the polymer chains of the individual particles diffuse into each other (interdiffusion) producing a coherent and continuous polymer film (stage 5). Studies carried out on polymer film formation have shown that there is also a level of polymer to be added to the cementitious material below which no film is formed. Microscopic analysis (scanning electron microscopy) carried out by (Afridi et al., 2003) on the microstructure of modified and unmodified cement pastes showed that this level is 5% latex polymer by weight of cement. In addition, latexes added to cementitious materials not only form the polymer film, they can interact physically and possibly chemically with the cementitious matrix to form the so-called co-matrix, which is a network structure in which the hydrated cement phase and the polymer phase interpenetrate. The following paragraph briefly describes the formation of this co-matrix.

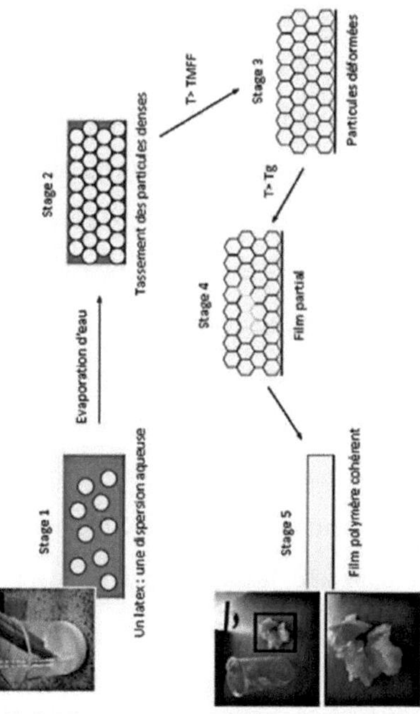

Figure 5. Schematic diagram of the polymer film formation mechanism.

d) Co-matrix formation mechanism :

Due to the wide variety of latexes, mineral binders and aggregates, as well as the complex mechanism of cement hydration, it is difficult to develop a valid general description for the formation of the microstructure of latex-modified mortars (Baueregger, 2014). A network structure is obtained, after the period of hardening and evaporation of the water from the mixture, where the cement hydrates and the polymer interpenetrate to form a co- matrix (Benali and Ghomari, 2017; Bahranifard, Vosoughi and Shariati, 2022). This co-matrix is constructed in several stages (Ngassam, 2013). The different stages of the model proposed by (Ohama, 1995), generally accepted and subsequently modified by (Gemert et al., 2005). Immediately after mixing, the cement particles and polymer particles are dispersed homogeneously in the cement paste. Initially, hydration of the cement takes place, resulting in an alkaline solution. In stage 1, some of the polymer particles are deposited on the surface of the anhydrous cement grains and aggregates. While, the other part can melt into a continuous film. This can partially or totally envelop the cement grains, and therefore inhibits setting and limits hydration of the cement grains.In the next stage, the cement hydrates and the polymer flocculates coalesce into a polymer film. The processes that take place in this stage depend on the curing conditions. If the dry curing period is not applied during maturation, the formation of the polymer film is delayed and the influence on the properties of the fresh mix is limited at this stage. On the other hand, if a dry curing period is added after wet curing, the hydrated structure develops and polymer film formation takes place. As the water in the mix evaporates and is consumed by the hydration of the cement, the polymer particles flocculate and form an organic layer on the surface of the unhydrated cement particles and adhere to the calcium silicate layer on the aggregates. Reactions also take place between the polymer, cement and aggregate particles. The end result is a continuous polymer film in which membranes bind the cement hydrates together to form a three-dimensional monolithic network. In this network, the polymeric phase interpenetrates the partially hydrated cementitious phase. As a result, the properties of ordinary cement mortars are greatly improved.

e) Cement-polymer interactions

When a latex-type polymer is added to a cementitious mixture, interactions can occur during cement hydration (Ngassam, 2013). The actual mechanisms of these interactions are not fully understood (Salbin, 1996; Soufi et al., 2015). Some research believes that only physical interactions exist between the two phases with the formation of the polymer film responsible for improving certain physical and mechanical properties (Soufi, 2013). Others believe that, in addition to the physical interactions, there are chemical interactions between the polymers and the cement hydrates responsible for the change in the microstructure of modified mortars (Ngassam, 2013; Soufi, 2013; Soufi et al., 2015).In a very recent study (Wang et al., 2016) investigated the modification mechanism of latex polymers on cement pastes. The authors used two types of latex one is based on Butyl benzene, this type was used to represent the group of polymers without active groups in their polymer chains. The second is a styrene-butadiene carboxylic latex that contains active groups capable of reacting with the hydration products to produce a 3D network structure in the modified systems. They found that, for the first latex without active groups, the mechanism responsible for the interaction was a physical one. The latex film, having covered the surfaces of the hydration crystals and filled the cracks and pores in the cement, is able to give better performance to the modified systems. In the case of the other polymer used, however, they have shown that the mechanism responsible includes both physical and chemical mechanisms. The physical modification mechanism was the same as that without active groups, however, in the chemical mechanism, the active groups react with the hydration products, binding the latex polymer chains together. These chemical reactions create a 3D network structure, which increases the flexural strength of the modified systems. On the other hand, according to (Yang et al., 2009), the added polymer absorbs the calcium formed and thus reduces the formation of Portlandite.The highly alkaline environment of fresh mortar ionises the small amount of carboxylic acid (which is chemically bound to the surface of the polymer particles), which tends to interact with the calcium ions produced by cement hydration, which subsequently results in better stability latex polymer and strong adhesion to existing substrates (Yang et al., 2009; Benali and Ghomari, 2017).These results are also confirmed by the work of (Wang et al., 2015) where they studied the chemical interaction mechanism in polyacrylate (PA) modified cement pastes. They found that the chemical reactions can be divided into three stages: the first concerns the cement hydration reaction. This reaction creates the high alkalinity of the mixture. In the second stage, the ester groups in the hydrolysed polyacrylate chains produce carboxylated groups in this highly alkaline environment. In the third stage, these carboxylated groups react with the cement's hydration products, Ca(OH)2, to generate a new Ca(HCOO)2 product. The chemical reaction between the latex and the cement generated a cross-linked network structure, in which Ca^{2+} acted as a cross-linking point and linked together the different polyacrylate chains. The result of this cross-linked network structure is the development of the cement's properties. (Tian et al., 2013), investigated the microstructure formation process of PA latex-modified systems. The authors found that PA particles react with calcium ions in the pores of the interstitial solution and cement hydrates. These PA polymers are combined and adsorbed onto the cement particles and hydrates forming C-type polymer-cement bonds. As a result, the PA polymer will not disperse uniformly in the fresh mix. (Larbi and Bijen, 1990), also studied the mechanism of these interactions. They used three types of latex polymer, one based on acrylic styrene, one based on acrylic styrene with a coupling agent, and one based on polyvinylidene chloride. The results show that these polymers react with Ca^{2+} , $S04^{2-}$, and OH^{-} , which are produced by the cement during hydration. A decrease was reported in the concentration of Ca^{2+} ions, against a high content of $S04^{2-}$ and OH^{-} in the interstitial solution of modified mortars compared with that of unmodified mortars. This shows that Ca^{2+} ions are trapped by the polymers, which in turn reduce the formation of $Ca(OH)_2$ portlandite and ettringite. These results are confirmed by other studies, such as the study carried out by (Afridi et al., 1989), where they found that the formation of $Ca(OH)_2$ in modified mortars is reduced probably due to the absorption of $Ca(OH)_2$ by the polymer film formed. This reduction depends on the P/C ratio and the type of polymer used. In this study polyvinyl carboxylase vinyl acetate, polyethylene vinyl acetate, and polyethylene vinyl acetate emulsions, have been shown to be more effective than styrene-butadiene latex in reducing the amount of Ca $(OH)_2$ in modified mortars. (Gomes and Ferreira, 2005), found that the addition of the copolymer (VA / VeoVa) greatly reduced the formation of Portlandite. (Betioli et al., 2009), in their study also found that the addition of the EVA polymer to cement paste reduces calcium hydroxide due to the chemical interaction between Ca^{2+} ions and acetate anions.

f) Binder-aggregate interactions :

According to (Ngassam, 2013), the binder-aggregate interface is the most fragile zone in mortars, due to the fact that it contains significant porosity and a large quantity of oriented portlandite crystals. Several researchers (Ohama, 1995; Sakai and Sugita, 1995; Yang et al., 2009) have shown that the reaction between Ca^{2+} ions and carboxyl groups in modified mortars greatly reduces the formation of portlandite and can lead to chemical bonds between the polymer and SiO_2 from the aggregate surface. This tends to reduce the porosity that exists in the binder-aggregate interface and bridges cracks, thereby increasing strength in this so-called fragile zone.

CONCLUSIONS

Incorporating polymers into concretes and mortars produces materials similar to ordinary cementitious concretes, but with superior characteristics. The use of concretes or mortars with synthetic and impregnated resins in the construction industry is currently limited by: the high cost of materials and specialised labour; application conditions (special temperature and humidity); and poor behaviour at high temperatures. Several types of polymer are used to modify portland cement-based mortars and concretes: latexes, redispensable polymers, water-soluble polymers and liquid resins. Latexes are the most widely used. They take the form of milky-white dispersions produced by emulsion polymerisation, which form polymer films after drying. They are used in relatively large quantities, ranging from 5 to 20% polymer solids by weight of the cement in the mix.Latex-modified mortars and concretes are easily prepared using conventional mixing equipment and tools. Curing in damp conditions such as water immersion or wet curing applicable to ordinary cement and concrete mortar is detrimental to latex-modified mortar and concrete. Optimal properties of modified systems are achieved by combined wet and dry curing. Latex-modified mortars are widely used rather than latex-modified concrete in terms of the balance between performance and cost.

REFERENCE

ACI_Committee_548 (1997) Guide for the Use of Polymers in Concrete.

ACI_Committee_548 (2003) Polymer-Modified Concrete.

ACI_Committee_548 (2009) Report on Polymer-Modified Concrete.

Adams, M., Browne, R. D. and French, E. L. (1975) 'Utilisation des concons de polymère', Batiment International, Building Research and Practice, 3(4), pp. 213- 231.

Afridi, M. U. . (1995) 'Water Retention and Adhesion of Powdered and Aqueous Polymer-Modified Mortars', Cement and Concrete Composites, 17, pp. 113-118.

Afridi, M. U. K. et al (1989) 'Behaviour of Ca(OH)2 in polymer modified mortars',

The International Journal of Cement Composites and Lightweight Concrete, 11(4),pp. 235-244.

Afridi, M. U. K. et al. (2003) 'Development of polymer films by the coalescence of polymer in powdered and aqueous polymer-modified mortars', Cement and Concrete Research, 33, pp. 1715 - 1721.

Aggarwal, L. K., Thapliyal, P. C. and Karade, S. R. (2007) 'Properties of polymer modified mortars using epoxy and acrylic emulsions', Construction and Building Materials, 21, pp. 379-383.

Amaral, M. D. (2004) 'Assessing the environmental cost of recent progresses in emulsion polymerization', Reactive & Functional Polymers, 58, pp. 197-202.

ASTM_C1439 (1999) Standard Test Methods for PolymerModified Mortar and Concrete.

Atsonios, K. et al. (2015) 'Integration of calcium looping technology in existing cement plant for CO2 capture: Process modeling and technical considerations',

Fuel, 153, pp. 210-223.

Bahranifard, Z., Vosoughi, A.-R. and Shariati, F. F. T. K. (2022) 'Effects of water- cement ratio and superplasticizer dosage on mechanical and microstructure formation of styrene-butyl acrylate copolymer concrete', Construction and Building Materials, 318, pp. 125889.

Barluenga, G. and Herna'ndez-Olivares, F. (2004) 'SBR latex modified mortar rheology and mechanical behaviour', Cement and Concrete Research, 34, pp. 527 - 535.

Baueregger, S. M. (2014) Interaction of Latex Polymers with Cement Based Building Materials. University of Munich Germany.

Belkhodja, A. (2013) Etude des interactions ciment - polymères dans un matériau de construction. Université Abou-Bekr Belkaid de Tlemcen.

Benali, Y. (2018) Etude du comportement mécanique et de durabilité des mortiers de polymères modifiés au latex. Université Aboubaker blkaid Tlemcen, Algeria.

Benali, Y. and Ghomari, F. (2017) 'Latex influence on the mechanical behavior and durability of cementitious materials', Journal of Adhesion Science and Technology,
31(3). doi: 10.1080/01694243.2016.1208378.

Benyahia, B. (2009) Modelling, experimentation and multi-criteria optimisation of an emulsion copolymerisation process in the presence of a chain transfer agent. University of Nancy.

Betioli, A. M. et al (2009) 'Chemical interaction between EVA and Portl and cement hydration at early-age', Construction and Building Materials, 23, pp. 3332-3336.

Bothra, S. R. and Ghugal, Y. M. (2015) 'Polymer -modified concrete: review',

International Journal of Research in Engineering and Technology, 04(04), pp. 845- 848.

Brien, J. V and Mahboub, K. C. (2013) 'Influence of polymer type on adhesion performance of a blended cement mortar', International Journal of Adhesion & Adhesives, 43, pp. 7-13.

Bur, N. (2012) Etude des caractéristiques physico-chimiques de nouveaux concrets éco-respectueux pour leur résistance à l'environnement dans le cadre du développement durable. University of Strasbourg.

Chetrashekhar, G. V, Cooper, E. I. and Shafer, M. W. (1989) 'Dielectric properties of macro-defect-free (M D F) cements', Journal of materials science, 24, pp. 3356-3360.

Czarnecki, L. (2007) 'Concrete- polymer Composites : Trends Shaping the Future',Int. J. Soc. Mater. Eng, 15(1), pp. 1-5.

Czarnecki, L., Őzkul, H. and Wang, R. (2013) 'Driving Forces Concrete Polymer Composites', Advanced Materials Research, 687, pp. 68-74.

Deventer, J. S. J. v, Provis, J. L. and Duxson, P. (2012) 'Technical and commercial progress in the adoption of geopolymer cement', Minerals Engineering, 29, pp. 89-104.

Dhaini, F. (2014) Study of latex cement model interactions. Consequences on rheological properties. University of Burgundy.

Donatello, S., Tyrer, M. and Cheeseman, C. R. (2009) 'Recent developments in macro-defect-free (MDF) cements', Construction and Building Materials, 23, pp. 1761-1767.

Elalaoui, O. (2012) Optimisation de la formulation et de la tenue aux hautes températures d'un béton à base d'époxyde. University of Tunis El Manar and University of Cergy-Pontoise (France).

Erdmenger, T. et al. (2010) 'Recent developments in the utilization of green solvents in polymer chemistry', Chemical Society Reviews, 39, pp. 3317-3333.

Fowler, D. W. (1999) 'Polymers in concrete: a vision for the 21st century', Cement & Concrete Composites, 21, pp. 449-452.

Fu, X. and Chung, D. D. L. (1996a) 'Effect of Methylcellulose admixture on the mechanical properties of cement', Cement and Concrete Research, 26(4), pp. 535-538.

Fu, X. and Chung, D. D. L. (1996b) 'Effect of polymer admixtures to cement on the bond strength and electrical contact resistivity between steel fibre and cement', Cement and Concrete Research, 26(2), pp. 189-194.

Fu, X., Lu, W. and Chung, D. D. L. (1996) 'Improving the bond strength between carbon fibre and cement by fibre surface treatment and polymer addition to cement mix', Cement and Concrete Research, 26(7), pp. 1007-1012.

Gemert, D. V et al. (2005) 'Cement concrete and concrete-polymer composites: Two merging worlds. A report from 11th ICPIC Congress in Berlin, 2004', Cement & Concrete Composites, 27, pp. 926-933.

Gomes, C. E. M. and Ferreira, O. P. (2005) 'Analyses of Microstructural Properties of VA/VeoVA Copolymer Modified Cement Pastes', Polímeros: Ciência e Tecnologia, 15(3), pp. 193-198.

Goto, T. (2006) Influence of latex molecular parameters on hydration, rheology and mechanical properties of cement/latex composites. University of Paris VI.

Haidar, M. (2011) Optimisation and durability of epoxy-based micro-concretes. University of Cergy-Pontoise.

Hasanbeigi, A., Price, L. and Lin, E. (2012) 'Emerging energy-efficiency and CO2 emission-reduction technologies for cement and concrete production: A technical review', Renewable and Sustainable Energy Reviews, 16, pp. 6220-6238.

Hot, J. (2013) Influence of superplasticizer polymers and air entraining agents on the macroscopic viscosity of cementitious materials. University of Paris-Est.

Jenni, A. et al. (2005) 'Influence of polymers on microstructure and adhesive strength of cementitious tile adhesive mortars', Cement and Concrete Research, 35, pp. 35-50.

Kapil Soni, E. and Joshi, Y. P. (2014) 'Performance Analysis of Styrene Butadiene Rubber-Latex on Cement Concrete Mixes', Journal of Engineering Research and Applications www.ijera.com, 4(3), pp. 838-844. Available at: www.ijera.com.

Kardon, J. B. (1997) 'Polymer-modified concrete: review', J. Mater. Civ. Eng, 9, pp. 85-92.

Kim, J.-H. and Robertson, R. E. (1997) 'Prevention of air void formation in polymer- modified cement mortar by pre-wetting', Cement and Concrete Research, 27(2), pp. 171-176.

Knapen, E. (2007) Microstructure formation in cement mortars modified with water-soluble polymers.

Knapen, E. and Gemert, D. Van (2015) 'Polymer film formation in cement mortars modified with water-soluble polymers', Cement & Concrete Composites, 58, pp. 23-28.

Larbi, J. A. and Bijen, J. M. J. M. (1990) 'Interaction of polymers with portland cement during hydration: a study of the chemistry of the pore solution of polymer modified cement systems', Cement and concrete research, 20, pp. 139-147.

Li, J., Zhong, S. and Zhang, C. (2012) 'Influence of superplasticiser and mixing procedure on properties of styrene-acrylic ester latex modified mortars', Magazine of Concrete Research, 64(5), pp. 411-417.

Li, P. et al. (2022) 'Time-dependent retardation effect of epoxy latexes on cement hydration: Experiments and multi component hydration model', Construction and

Building Materials, 320, pp. 126282.

Ngassam, I. L. T. (2013) Durability of repairs to concrete engineering structures. University of Paris-Est.

Nguyen, D. (2008) Etude de la nucleation contrôlée de latex polymère à la surface de nanoparticules d'oxyde pour l'élaboration de colloides hybrides structures. University of Bordeaux 1.

Nicot, P. (2008) 'Mortar-substrate interactions: determining factors in mortar performance and adhesion', p. 224.

Ohama, Y. (1995) Handbook of polymer- modified concrete and mortars Properties and Process Technology. Noyes Publications Mill Road, Park Ridge, New Jersey 07656.

Ohama, Y. (1998) 'Polymer-based Admixtures." Cement and Concrete Composites', Cement and Concrete Composites, 20, pp. 189-212.

Ohama, Y. (2011) 'Concrete-Polymer Composites - The Past, Present and Future -', Key Engineering Materials, 466, pp. 1-14.

Pragliola, S., Vita, R. D. and Longo, P. (2013) 'Aqueous emulsion polymerization of styrene and substituted styrenes using titanocene compounds', Polymer, 54, pp. 1583-1587.

Ribeiro, M. C. dos S. (2006) New Polymer Mortar Formulations - Development, Characterization and Application Forms -. Faculty of Engineering of University of Porto.

Sakai, E. and Sugita, J. (1995) 'Composite mechanism of polymer modified cement', Cement and Concrete Research, 25(1), pp. 127-135.

Salbin, M. K. (1996) Properties and performance of polymer modified cements and mortars. University of Aston in Birmingham.

Sivakumar, M. V. N. (2011) 'Effect of Polymer modification on mechanical and structural properties of concrete An experimental investigation', International journal of civil and structural engineering, 1(4), pp. 732-740.

Soufi, A. (2013) Étude de la durabilité des systèmes béton armé - mortiers de réparation en milieu marin. University of La Rochelle.

Soufi, A. et al. (2015) 'Influence of polymer proportion on transfer properties of repair mortars having equivalent water porosity', Materials and Structures, 49(1), pp. 383-398.

Thickett, S. C. and Gilbert, R. G. (2007) 'Emulsion polymerization: State of the art in kinetics and mechanisms', Polymer, 48, pp. 6965-6991.

Tian, Y. et al. (2013) 'Research on the microstructure formation of polyacrylate latex modified mortars', Construction and Building Materials, 47, pp. 1381-1394.

Ukrainczyk, N. and Rogina, A. (2013) 'Styrene-butadiene latex modified calcium aluminate cement mortar', Cement & Concrete Composites, 41, pp. 16-23.

Wang, M. et al. (2015) 'Research on the chemical mechanism in the polyacrylate latex modified cement system', Cement and Concrete Research, 76, pp. 62-69.

Wang, M. et al. (2016) 'Research on the mechanism of polymer latex modified cement', Construction and Building Materials, 111, pp. 710-718.

Yang, Z. et al. (2009) 'Effect of styrene- butadiene rubber latex on the chloride permeability and microstructure of Portlet cement mortar', Construction and Building Materials, (23), pp. 2283-2290.

Youssef, I. (2012) Emulsion and miniemulsion polymerization. Influence of the combination of molecular and macromolecular stabilizers and in situ monitoring by Raman spectroscopy. University of Lorraine.

Zabihi, N. and Ozkul, M. H. (2015) 'The Effect of Nano-Silica Particles on Fresh and Hardened State Properties of Polymer Cement Mortars', Advanced

Materials Research, 1129, pp. 113-120.

Zhang, W. (2014) Synthesis and fracture toughness of macro-defect-free (MDF) cement. University of Illinois at Urbana-Champaign.

Zohhadi, N. (2014) Bio-Inspired and Low-Content Polymer Cement Mortar for Structural Rehabilitation. University of South Carolina.

Printed by Books on Demand GmbH, Norderstedt / Germany